C000194089

"Exercises of Double and Triple Integrals"

SIMONE MALACRIDA

In this book, exercises are carried out regarding the following mathematical topics:
double integrals
triple integrals
Initial theoretical hints are also presented to make the performance of the exercises understandable

Simone Malacrida (1977)
Engineer and writer, has worked on research,
finance, energy policy and industrial plants.

AUTHOR'S NOTE
To understand what is reported in this manual, it is necessary to have knowledge of the theory of mathematical analysis at university level.

# ANALYTICAL INDEX

# INTRODUCTION

In this exercise book, some examples of calculations relating to double and triple integrals are carried out.

These integrals represent the most used operations for real functions with several variables especially in physics and technology.

In order to understand in more detail what is explained in the resolution of the exercises, the reference theoretical context is recalled at the beginning of each chapter.

What is exposed in this workbook is generally addressed in advanced mathematical analysis courses (analysis 2) and, as such, a knowledge of at least the main properties of real functions with several variables is required, such as the concepts of mixed derivatives, differentiability and the Jacobian formalism .

# Introduction

I

# DOUBLE INTEGRALS

Given a function of several real variables, it is possible to define **the multiple integral according to Riemann** , ie the integral carried out in each of its variables.

The theorems of the integral mean and the weighted mean as well as the properties of monotonicity,additivity and linearity always hold.

The notation is the same except to make explicit the writing for double and triple integrals:

$$\iint_T f(x, y)\, dx\, dy$$

$$\iiint_S f(x, y, z)\, dx\, dy\, dz$$

Similarly, improper integrals maintain their properties and the methods of solving a multiple integral are the same, almost always accompanied by attempts to reduce the multiple integral into n

# I – Double integrals

different and successive integrations with respect to the n single variables.

**The reduction formulas are based on the assumption of a corollary of Fubini's theorem according to which if the integrand function, for example of two real variables, can be expressed as the product of two functions defined on a single variable, then also the integral double can be reduced to the product of two simple integrals:**

$$f(x, y) = h(x)g(y) \Rightarrow$$
$$\iint_T f(x, y)d(x, y) = \iint_T h(x)g(y)dxdy = \left( \int_A h(x)dx \right)\left( \int_B g(y)dy \right)$$

Where the set T is given by the Cartesian product of the sets A and B.

This corollary guarantees this result for functions where the integral of the absolute value converges; Tonelli's theorem guarantees the same result for positive functions.

Such a reduction is also possible if the integrand function is continuous and the integration set is bounded.

If a change of variable is made in the calculation of multiple integrals, the following relation holds:

$$\int_A f(x_1,...,x_n)d(x_1,...,x_n) = \int_B f(g(y_1,...,y_n))\left|\det(J_g(y))\right|d(y_1,...,y_n)$$

## I – Double integrals

The determinant of the Jacobian matrix relating to the function after the change of variable appears in the formula; in polar and cylindrical coordinates this determinant is equal to r, in three-dimensional spherical coordinates it is equal to $r^2 \sin \vartheta$.

## **_Exercise 1_**

Calculate the following double integral over the specified set:

$$\int_\Omega (x + y)\, dx\, dy. \qquad \Omega = \left\{ (x, y) \in \mathbb{R}^2 : \ 0 < y < \frac{\sqrt{2}}{2},\ y < x < \sqrt{1 - y^2} \right\}$$

The given set is x-simple so we can safely separate the variables and integrate:

$$\int_\Omega (x + y)\, dx\, dy = \int_0^{\frac{\sqrt{2}}{2}} \left[ \int_y^{\sqrt{1-y^2}} (x + y)\, dx \right] dy =$$

$$= \int_0^{\frac{\sqrt{2}}{2}} \left[ \frac{1}{2} x^2 + xy \right]_y^{\sqrt{1-y^2}} dy = \int_0^{\frac{\sqrt{2}}{2}} \left[ \frac{1}{2}\left(1 - y^2\right) + y\sqrt{1 - y^2} - \frac{3}{2} y^2 \right] dy =$$

$$= \int_0^{\frac{\sqrt{2}}{2}} \left( \frac{1}{2} - 2y^2 + y\sqrt{1 - y^2} \right) dy = \left[ \frac{1}{2} y - \frac{2}{3} y^3 - \frac{1}{3}\left(1 - y^2\right)^{\frac{3}{2}} \right]_0^{\frac{\sqrt{2}}{2}} = \frac{1}{3}.$$

# I – Double integrals

## *Exercise 2*

Calculate the following double integral over the specified set:

$$\int_{\Omega} \left( x^2 + y^2 \right) dx\, dy, \qquad \Omega = \left\{ (x,y) \in \mathbb{R}^2 : \ 0 \le x \le 1, \ 1 \le y \le 2 \right\}$$

The given set is x-simple and y-simple and so we can separate the variables and integrate:

$$\int_{\Omega} \left( x^2 + y^2 \right) dx\, dy = \int_0^1 \left[ \int_1^2 \left( x^2 + y^2 \right) dy \right] dx =$$

$$= \int_0^1 \left[ x^2 y + \frac{1}{3} y^3 \right]_1^2 dx = \int_0^1 \left( x^2 + \frac{7}{3} \right) dx = \left[ \frac{1}{3} x^3 + \frac{7}{3} x \right]_0^1 = \frac{8}{3}.$$

## *Exercise 3*

Calculate the following double integral over the specified set:

$$\int_{\Omega} xy\, dx\, dy, \qquad \Omega = \left\{ (x,y) \in \mathbb{R}^2 : \ 0 < x < 1, \ x^2 < y < \sqrt{x} \right\}$$

The given set is y-simple so we can safely separate the variables and integrate:

# I – Double integrals

$$\int_\Omega xy\,dx\,dy = \int_0^1 \left[ \int_{x^2}^{\sqrt{x}} xy\,dy \right] dx =$$

$$= \int_0^1 x \left[ \frac{1}{2}y^2 \right]_{x^2}^{\sqrt{x}} dx = \frac{1}{2}\int_0^1 \left( x^2 - x^5 \right) dx = \frac{1}{2}\left[ \frac{1}{3}x^3 - \frac{1}{6}x^6 \right]_0^1 = \frac{1}{12}.$$

## *Exercise 4*

Calculate the following double integral over the specified set:

$$\int_\Omega \frac{xy}{x^2+y^2}\,dx\,dy, \qquad \Omega = \left\{ (x,y) \in \mathbb{R}^2 : \ 1 < x^2+y^2 < 4, \ x > 0, \ y > 0 \right\}$$

The given set is x-simple and y-simple, moreover it has radial symmetry and therefore we pass in polar coordinates:

$$\Phi : \begin{cases} x = \rho\cos\vartheta \\ y = \rho\sin\vartheta, \end{cases} \quad \rho \geq 0, \ 0 \leq \vartheta \leq 2\pi, \quad |\det J_\Phi(\rho,\vartheta)| = \rho.$$

From which it follows that:

$$(x,y) \in \Omega \quad \Longleftrightarrow \quad \begin{cases} 1 < \rho < 2 \\ 0 < \vartheta < \frac{\pi}{2}. \end{cases}$$

# I – Double integrals

$$\Omega' = \left\{ (\rho, \vartheta) \in \mathbb{R}^2 : \ 1 < \rho < 2, \ 0 < \vartheta < \frac{\pi}{2} \right\}.$$

Substituting, we have:

$$\int_\Omega \frac{xy}{x^2 + y^2} \, dx \, dy = \int_{\Omega'} \rho \cos \vartheta \sin \vartheta \, d\rho \, d\vartheta =$$

$$= \left( \int_1^2 \rho \, d\rho \right) \left( \int_0^{\frac{\pi}{2}} \cos \vartheta \sin \vartheta \, d\vartheta \right) = \left[ \frac{1}{2} \rho^2 \right]_1^2 \left[ \frac{1}{2} \sin^2 \vartheta \right]_0^{\frac{\pi}{2}} = \frac{3}{4}.$$

## *Exercise 5*

Calculate the following double integral over the specified set:

$$\int_\Omega xy \, dx \, dy, \qquad \Omega = \left\{ (x, y) \in \mathbb{R}^2 : \ x^2 + y^2 < 1, \ x^2 + y^2 < 2x, \ y > 0 \right\}$$

Let's go to polar coordinates:

$$\Phi : \begin{cases} x = \rho \cos \vartheta \\ y = \rho \sin \vartheta, \end{cases} \quad \rho \geq 0, \ 0 \leq \vartheta \leq 2\pi, \quad |\det J_\Phi(\rho, \vartheta)| = \rho.$$

From which it follows that:

# I – Double integrals

$$(x, y) \in \Omega \quad \Longleftrightarrow \quad \begin{cases} 0 < \rho < 1 \\ 0 < \rho < 2\cos\vartheta \\ 0 < \vartheta < \frac{\pi}{2}. \end{cases}$$

$$\Omega' = \Omega'_1 \cup \Omega'_2$$

$$\Omega'_1 = \left\{ (\rho, \vartheta) \in \mathbb{R}^2 : \ 0 < \rho < 1, \ 0 < \vartheta < \frac{\pi}{3} \right\},$$

$$\Omega'_2 = \left\{ (\rho, \vartheta) \in \mathbb{R}^2 : \ 0 < \rho < 2\cos\vartheta, \ \frac{\pi}{3} \leq \vartheta < \frac{\pi}{2} \right\}.$$

Substituting, we have:

$$\int_\Omega xy \, dx \, dy = \int_{\Omega'} \rho^3 \cos\vartheta \sin\vartheta \, d\rho \, d\vartheta =$$

$$= \int_{\Omega'_1} \rho^3 \cos\vartheta \sin\vartheta \, d\rho \, d\vartheta + \int_{\Omega'_2} \rho^3 \cos\vartheta \sin\vartheta \, d\rho \, d\vartheta =$$

$$= \left( \int_0^1 \rho^3 \, d\rho \right) \left( \int_0^{\frac{\pi}{3}} \cos\vartheta \sin\vartheta \, d\vartheta \right) + \int_{\frac{\pi}{3}}^{\frac{\pi}{2}} \cos\vartheta \sin\vartheta \left[ \int_0^{2\cos\vartheta} \rho^3 \, d\rho \right] d\vartheta =$$

$$= \left[ \frac{1}{4}\rho^4 \right]_0^1 \left[ \frac{1}{2}\sin^2\vartheta \right]_0^{\frac{\pi}{3}} + \int_{\frac{\pi}{3}}^{\frac{\pi}{2}} \cos\vartheta \sin\vartheta \left[ \frac{1}{4}\rho^4 \right]_0^{2\cos\vartheta} d\vartheta =$$

$$= \frac{3}{32} + 4 \int_{\frac{\pi}{3}}^{\frac{\pi}{2}} \cos^5\vartheta \sin\vartheta \, d\vartheta = \frac{3}{32} + 4 \left[ -\frac{1}{6}\cos^6\vartheta \right]_{\frac{\pi}{3}}^{\frac{\pi}{2}} = \frac{5}{48}.$$

I – Double integrals

# *Exercise 6*

Calculate the following double integral over the specified set:

$$\int_\Omega xy\,dx\,dy, \qquad \Omega = \left\{(x,y) \in \mathbb{R}^2 : x^2 + 2y^2 < 1,\ x > 0,\ y > 0\right\}$$

Let's switch to elliptical coordinates:

$$\Phi : \begin{cases} x = \rho\cos\vartheta \\ y = \frac{\sqrt{2}}{2}\rho\sin\vartheta, \end{cases} \quad \rho \geq 0,\ 0 \leq \vartheta \leq 2\pi, \quad |\det J_\Phi(\rho,\vartheta)| = \frac{\sqrt{2}}{2}\rho.$$

From which it follows that:

$$(x,y) \in \Omega \quad \Longleftrightarrow \quad \begin{cases} 0 < \rho < 1 \\ 0 < \vartheta < \frac{\pi}{2}. \end{cases}$$

$$\Omega' = \left\{(\rho,\vartheta) \in \mathbb{R}^2 : 0 < \rho < 1,\ 0 < \vartheta < \frac{\pi}{2}\right\}.$$

Substituting, we have:

$$\int_\Omega xy\,dx\,dy = \frac{1}{2}\int_{\Omega'} \rho^3 \cos\vartheta \sin\vartheta\,d\rho\,d\vartheta =$$

# I – Double integrals

$$= \frac{1}{2} \left( \int_0^1 \rho^3 \, d\rho \right) \left( \int_0^{\frac{\pi}{2}} \cos \vartheta \sin \vartheta \, d\vartheta \right) = \frac{1}{2} \left[ \frac{1}{4} \rho^4 \right]_0^1 \left[ \frac{1}{2} \sin^2 \vartheta \right]_0^{\frac{\pi}{2}} = \frac{1}{16}.$$

## Exercise 7

Calculate the following double integral over the specified set:

$$\int_\Omega x(1-y) \, dx \, dy, \qquad \Omega = \left\{ (x,y) \in \mathbb{R}^2 : \ 0 < y < \frac{\sqrt{2}}{2}, \ y < x < \sqrt{1-y^2} \right\}$$

The set is x-simple and therefore it is easy to obtain that:

$$\int_\Omega x(1-y) \, dx \, dy = \int_0^{\frac{\sqrt{2}}{2}} \left[ \int_y^{\sqrt{1-y^2}} x(1-y) \, dx \right] dy =$$

$$= \int_0^{\frac{\sqrt{2}}{2}} \left[ \frac{1}{2} x^2 (1-y) \right]_y^{\sqrt{1-y^2}} dy = \frac{1}{2} \int_0^{\frac{\sqrt{2}}{2}} (1-y) \left( 1 - 2y^2 \right) dy =$$

$$= \frac{1}{2} \int_0^{\frac{\sqrt{2}}{2}} \left( 1 - y - 2y^2 + 2y^3 \right) dy = \frac{1}{2} \left[ y - \frac{1}{2} y^2 - \frac{2}{3} y^3 + \frac{1}{2} y^4 \right]_0^{\frac{\sqrt{2}}{2}} = \frac{\sqrt{2}}{6} - \frac{1}{16}.$$

## Exercise 8

Calculate the following double integral over the specified set:

# I – Double integrals

$$\int_\Omega \log(xy)\, dx\, dy. \qquad \Omega = \left\{ (x,y) \in \mathbb{R}^2 : \ -1 < x < -\frac{1}{2}, \ 4x < y < \frac{1}{x} \right\}$$

The set is y-simple and therefore it is easy to obtain that:

$$\int_\Omega \log(xy)\, dx\, dy = \int_{-1}^{-\frac{1}{2}} \left[ \int_{4x}^{\frac{1}{x}} \log(xy)\, dy \right] dx =$$

We integrate by parts:

$$= \int_{-1}^{-\frac{1}{2}} \left( \left[ y \log(xy) \right]_{4x}^{\frac{1}{x}} - \int_{4x}^{\frac{1}{x}} dy \right) dx = \int_{-1}^{-\frac{1}{2}} \left( -4x \log 4x^2 - \frac{1}{x} + 4x \right) dx =$$

Further integrating by parts, we have:

$$= -4 \left( \left[ \frac{1}{2} x^2 \log 4x^2 \right]_{-1}^{-\frac{1}{2}} - \int_{-1}^{-\frac{1}{2}} x\, dx \right) + \left[ -\log |x| + 2x^2 \right]_{-1}^{-\frac{1}{2}} =$$

$$= -4 \left( -\log 2 + \frac{3}{8} \right) + \log 2 - \frac{3}{2} = 5 \log 2 - 3.$$

## **_Exercise 9_**

Calculate the following double integral over the specified set:

# I – Double integrals

$$\int_{\Omega} \log \frac{x}{y^2}\, dx\, dy. \qquad \Omega = \left\{ (x,y) \in \mathbb{R}^2 : \frac{1}{4}x < y^2 < x,\ 1 < xy < 2 \right\}$$

The given set can be decomposed into a finite number of x-simple or y-simple sets having two by two in common at most segments of the plane.
Being contained in the first quadrant, we have:

$$(x,y) \in \Omega \quad \Longleftrightarrow \quad \begin{cases} \frac{1}{4} < \frac{y^2}{x} < 1, \\ 1 < xy < 2. \end{cases}$$

We make the following change of variables:

$$\Psi : \begin{cases} u = xy \\ v = \frac{x}{y^2}, \end{cases} \quad (x,y) \in \Omega.$$

It follows that:

$$(x,y) \in \Omega \quad \Longleftrightarrow \quad \begin{cases} 1 < u < 2, \\ 1 < v < 4. \end{cases}$$

13

# I – Double integrals

$$\Phi : \begin{cases} x = u \sqrt[3]{\dfrac{v}{u}} \\[2mm] y = \sqrt[3]{\dfrac{u}{v}}, \end{cases} \qquad 1 < u < 2, \ 1 < v < 4.$$

From the local inversion theorem we have:

$$\det J_\Phi(u, v) = \det J_{\Psi^{-1}}(u, v) = \Big(\det J_\Psi(\Phi(u, v))\Big)^{-1}.$$

Considering that:

$$J_\Psi(x, y) = \begin{pmatrix} \frac{\partial u}{\partial x}(x, y) & \frac{\partial u}{\partial y}(x, y) \\ \frac{\partial v}{\partial x}(x, y) & \frac{\partial v}{\partial y}(x, y) \end{pmatrix} = \begin{pmatrix} y & x \\ \frac{1}{y^2} & -\frac{2x}{y^3} \end{pmatrix} \implies \det J_\Psi(x, y) = -\frac{3x}{y^2}.$$

It obviously follows:

$$\big| \det J_\Phi(u, v) \big| = \frac{1}{3v}.$$

$$\Omega' = \Big\{(u, v) \in \mathbb{R}^2 : \ 1 < u < 2, \ 1 < v < 4\Big\}.$$

Substituting, we have:

$$\int_\Omega \log \frac{x}{y^2}\, dx\, dy = \int_{\Omega'} \frac{\log v}{3v}\, du\, dv =$$

$$= \frac{1}{3} \left( \int_1^2 du \right) \left( \int_1^4 \frac{\log v}{v} \, dv \right) = \frac{1}{3} \Big[ u \Big]_1^2 \left[ \frac{1}{2} \log^2 v \right]_1^4 = \frac{1}{6} \log^2 4.$$

## *Exercise 10*

Calculate the following double integral over the specified set:

$$\int_\Omega \frac{1}{(x+y)^2} \, dx \, dy, \qquad \Omega = \Big\{ (x,y) \in \mathbb{R}^2 : \ 1 \le x \le 2, \ 3 \le y \le 4 \Big\}$$

The set is both x-simple and y-simple.
We therefore have:

$$\int_\Omega \frac{1}{(x+y)^2} \, dx \, dy = \int_1^2 \left[ \int_3^4 \frac{1}{(x+y)^2} \, dy \right] dx = \int_1^2 \left[ -\frac{1}{x+y} \right]_3^4 dx =$$

$$= \int_1^2 \left( \frac{1}{x+3} - \frac{1}{x+4} \right) dx = \Big[ \log|x+3| - \log|x+4| \Big]_1^2 = \log 25 - \log 24.$$

## *Exercise 11*

Calculate the following double integral over the specified set:

$$\int_\Omega \frac{x}{x^2 + y^2} \, dx \, dy, \qquad \Omega = \Big\{ (x,y) \in \mathbb{R}^2 : \ x^2 < y < 2x^2, \ 1 < x < 2 \Big\}$$

# I – Double integrals

The set is y-simple.
Therefore:

$$\int_\Omega \frac{x}{x^2+y^2}\,dx\,dy = \int_1^2 \left[\int_{x^2}^{2x^2} \frac{x}{x^2+y^2}\,dy\right]dx = \int_1^2\left[\int_{x^2}^{2x^2} \frac{\frac{1}{x}}{1+\left(\frac{y}{x}\right)^2}\,dy\right]dx =$$

$$= \int_1^2 \left[\arctan\left(\frac{y}{x}\right)\right]_{x^2}^{2x^2}\,dx = \int_1^2 (\arctan 2x - \arctan x)\,dx =$$

Integrating by parts, we get the result:

$$= \left[x(\arctan 2x - \arctan x)\right]_1^2 - \int_1^2 \left(\frac{2x}{1+4x^2} - \frac{x}{1+x^2}\right)dx =$$

$$= 2\arctan 4 - 3\arctan 2 + \frac{\pi}{4} - \left[\frac{1}{4}\log\left(1+4x^2\right) - \frac{1}{2}\log\left(1+x^2\right)\right]_1^2 =$$

$$= 2\arctan 4 - 3\arctan 2 + \frac{\pi}{4} - \frac{1}{4}\log 17 + \frac{3}{4}\log 5 - \frac{1}{2}\log 2.$$

# *Exercise 12*

Calculate the following double integral over the specified set:

$$\int_\Omega \frac{\sin y^2}{y}\,dx\,dy, \qquad \Omega = \left\{(x,y) \in \mathbb{R}^2 : \ 0 < x < y^2, \ \sqrt{\pi} < y < \sqrt{2\pi}\right\}$$

The set is x-simple, so we have:

## I – Double integrals

$$\int_{\Omega} \frac{\sin y^2}{y}\, dx\, dy = \int_{\sqrt{\pi}}^{\sqrt{2\pi}} \left[ \int_0^{y^2} \frac{\sin y^2}{y}\, dx \right] dy = \int_{\sqrt{\pi}}^{\sqrt{2\pi}} \left[ \frac{\sin y^2}{y} x \right]_0^{y^2} dy =$$

$$= \int_{\sqrt{\pi}}^{\sqrt{2\pi}} y \sin y^2\, dy = \left[ -\frac{1}{2} \cos y^2 \right]_{\sqrt{\pi}}^{\sqrt{2\pi}} = -1.$$

## *Exercise 13*

Calculate the following double integral over the specified set:

$$\int_{\Omega} xy\, dx\, dy, \qquad \Omega = \left\{ (x,y) \in \mathbb{R}^2 : \ x^2 + 2y^2 < 1 \right\}$$

The function and the set have symmetry about both axes.

This means that:

$$(x,y) \in \Omega,\ y > 0 \quad \Longrightarrow \quad (x,-y) \in \Omega, \quad f(x,-y) = -f(x,y).$$

Therefore:

$$\int_{\Omega \cap \{(x,y):\ y \geq 0\}} xy\, dx\, dy = -\int_{\Omega \cap \{(x,y):\ y < 0\}} xy\, dx\, dy.$$

Since the set is given by:

$$\Omega = \left( \Omega \cap \left\{ (x,y) \in \mathbb{R}^2 : \ y \geq 0 \right\} \right) \cup \left( \Omega \cap \left\{ (x,y) \in \mathbb{R}^2 : \ y < 0 \right\} \right)$$

Evidently the integral is zero, i.e.:

$$\int_{\Omega} xy \, dx \, dy = 0.$$

## *Exercise 14*

Calculate the following double integral over the specified set:

$$\int_{\Omega} \sqrt{x^2 + y^2} \, dx \, dy, \qquad \Omega = \left\{ (x, y) \in \mathbb{R}^2 : \; x^2 + y^2 - 4x < 0 \right\}$$

Let's go to polar coordinates:

$$\Phi : \begin{cases} x = \rho \cos \vartheta \\ y = \rho \sin \vartheta, \end{cases} \quad \rho \geq 0, \; -\pi \leq \vartheta \leq \pi, \quad |\det J_\Phi(\rho, \vartheta)| = \rho.$$

It follows that:

$$(x, y) \in \Omega \quad \Longleftrightarrow \quad \begin{cases} 0 < \rho < 4 \cos \vartheta \\ -\frac{\pi}{2} < \vartheta < \frac{\pi}{2}. \end{cases}$$

$$\Omega' = \left\{ (\rho, \vartheta) \in \mathbb{R}^2 : \; 0 < \rho < 4 \cos \vartheta, \; -\frac{\pi}{2} < \vartheta < \frac{\pi}{2} \right\}$$

# I – Double integrals

Substituting, we have:

$$\int_\Omega \sqrt{x^2 + y^2}\, dx\, dy = \int_{\Omega'} \rho^2\, d\rho\, d\vartheta =$$

$$= \int_{-\frac{\pi}{2}}^{\frac{\pi}{2}} \left( \int_0^{4\cos\vartheta} \rho^2\, d\rho \right) d\vartheta = \int_{-\frac{\pi}{2}}^{\frac{\pi}{2}} \left[ \frac{1}{3}\rho^3 \right]_0^{4\cos\vartheta} d\vartheta = \frac{64}{3} \int_{-\frac{\pi}{2}}^{\frac{\pi}{2}} \cos^3\vartheta\, d\vartheta =$$

Integrating by parts, we get the result:

$$= \frac{64}{3} \left( \left[ \sin\vartheta \cos^2\vartheta \right]_{-\frac{\pi}{2}}^{\frac{\pi}{2}} + 2 \int_{-\frac{\pi}{2}}^{\frac{\pi}{2}} \cos\vartheta \sin^2\vartheta\, d\vartheta \right) = \frac{128}{3} \left[ \frac{1}{3}\sin^3\vartheta \right]_{-\frac{\pi}{2}}^{\frac{\pi}{2}} = \frac{256}{9}.$$

## *Exercise 15*

Calculate the following double integral over the specified set:

$$\int_\Omega \left( x + y^2 \right) dx\, dy, \qquad \Omega = \left\{ (x, y) \in \mathbb{R}^2 : \ 1 < x^2 + y^2 < 4, \ x > 0, \ y > 0 \right\}$$

Let's go to polar coordinates:

$$\Phi : \begin{cases} x = \rho\cos\vartheta \\ y = \rho\sin\vartheta, \end{cases} \quad \rho \geq 0, \ -\pi \leq \vartheta \leq \pi, \quad |\det J_\Phi(\rho, \vartheta)| = \rho.$$

It follows that:

# I – Double integrals

$$(x, y) \in \Omega \quad \Longleftrightarrow \quad \begin{cases} 1 < \rho < 2 \\ 0 < \vartheta < \frac{\pi}{2}. \end{cases}$$

$$\Omega' = \left\{ (\rho, \vartheta) \in \mathbb{R}^2 : \ 1 < \rho < 2, \ 0 < \vartheta < \frac{\pi}{2} \right\}.$$

Substituting, we have:

$$\int_{\Omega} \left( x + y^2 \right) dx \, dy = \int_{\Omega'} \left( \rho^2 \cos \vartheta + \rho^3 \sin^2 \vartheta \right) d\rho \, d\vartheta =$$

$$= \left( \int_1^2 \rho^2 \, d\rho \right) \left( \int_0^{\frac{\pi}{2}} \cos \vartheta \, d\vartheta \right) + \left( \int_1^2 \rho^3 \, d\rho \right) \left( \int_0^{\frac{\pi}{2}} \sin^2 \vartheta \, d\vartheta \right) =$$

$$= \left[ \frac{1}{3} \rho^3 \right]_1^2 \left[ \sin \vartheta \right]_0^{\frac{\pi}{2}} + \left[ \frac{1}{4} \rho^4 \right]_1^2 \left[ \frac{1}{2} \left( \vartheta - \sin \vartheta \cos \vartheta \right) \right]_0^{\frac{\pi}{2}} = \frac{7}{3} + \frac{15}{16} \pi.$$

## Exercise 16

Calculate the following double integral over the specified set:

$$\int_{\Omega} x \sqrt{x^2 + y^2} \, dx \, dy. \qquad \Omega = \left\{ (x, y) \in \mathbb{R}^2 : \ x^2 + y^2 < 1, \ x^2 + y^2 < 2y, \ x < 0 \right\}$$

Let's go to polar coordinates:

# I – Double integrals

$$\Phi : \begin{cases} x = \rho\cos\vartheta \\ y = \rho\sin\vartheta, \end{cases} \quad \rho \geq 0, \ -\pi \leq \vartheta \leq \pi, \quad |\det J_\Phi(\rho,\vartheta)| = \rho.$$

It follows that:

$$(x,y) \in \Omega \quad \Longleftrightarrow \quad \begin{cases} 0 < \rho < 1 \\ 0 < \rho < 2\sin\vartheta \\ \frac{\pi}{2} < \vartheta < \pi. \end{cases}$$

$$\Omega' = \Omega'_1 \cup \Omega'_2$$

$$\Omega'_1 = \left\{ (\rho,\vartheta) \in \mathbb{R}^2 : \ 0 < \rho < 1, \ \frac{\pi}{2} < \vartheta < \frac{5}{6}\pi \right\},$$

$$\Omega'_2 = \left\{ (\rho,\vartheta) \in \mathbb{R}^2 : \ 0 < \rho < 2\sin\vartheta, \ \frac{5}{6}\pi \leq \vartheta < \pi \right\}.$$

Substituting, we have:

$$\int_\Omega x\sqrt{x^2+y^2}\,dx\,dy = \int_{\Omega'} \rho^3\cos\vartheta\,d\rho\,d\vartheta =$$

$$= \int_{\Omega'_1} \rho^3\cos\vartheta\,d\rho\,d\vartheta + \int_{\Omega'_2} \rho^3\cos\vartheta\,d\rho\,d\vartheta =$$

$$= \left( \int_0^1 \rho^3\,d\rho \right)\left( \int_{\frac{\pi}{2}}^{\frac{5}{6}\pi} \cos\vartheta\,d\vartheta \right) + \int_{\frac{5}{6}\pi}^{\pi} \cos\vartheta \left[ \int_0^{2\sin\vartheta} \rho^3\,d\rho \right] d\vartheta =$$

$$= \left[\frac{1}{4}\rho^4\right]_0^1 \left[\sin\vartheta\right]_{\frac{\pi}{2}}^{\frac{5}{6}\pi} + \int_{\frac{5}{6}\pi}^{\pi} \cos\vartheta \left[\frac{1}{4}\rho^4\right]_0^{2\sin\vartheta} d\vartheta =$$

$$= -\frac{1}{8} + 4\int_{\frac{5}{6}\pi}^{\pi} \cos\vartheta \sin^4\vartheta \, d\vartheta = -\frac{1}{8} + 4\left[\frac{1}{5}\sin^5\vartheta\right]_{\frac{5}{6}\pi}^{\pi} =$$

$$= -\frac{1}{8} - \frac{1}{40} = -\frac{3}{20}.$$

## *Exercise 17*

Calculate the following double integral over the specified set:

$$\int_\Omega (x+y)\, dx\, dy. \qquad \Omega = \left\{(x,y) \in \mathbb{R}^2 : \; 2x^2 + 3y^2 < 4, \; x > 0, \; y > 0\right\}$$

Let's switch to elliptical coordinates:

$$\Phi : \begin{cases} x = \sqrt{2}\rho\cos\vartheta \\ y = \frac{2}{3}\sqrt{3}\rho\sin\vartheta, \end{cases} \quad \rho \geq 0, \; 0 \leq \vartheta \leq 2\pi, \quad |\det J_\Phi(\rho,\vartheta)| = \frac{2}{3}\sqrt{6}\rho.$$

It follows that:

$$(x,y) \in \Omega \quad \Longleftrightarrow \quad \begin{cases} 0 < \rho < 1 \\ 0 < \vartheta < \frac{\pi}{2}. \end{cases}$$

$$\Omega' = \left\{(\rho,\vartheta) \in \mathbb{R}^2 : \; 0 < \rho < 1, \; 0 < \vartheta < \frac{\pi}{2}\right\}.$$

# I – Double integrals

## Substituting, we have:

$$\int_{\Omega} (x+y)\, dx\, dy = \int_{\Omega'} \frac{2}{3}\sqrt{6}\rho\left(\sqrt{2}\rho\cos\vartheta + \frac{2}{3}\sqrt{3}\rho\sin\vartheta\right)\, d\rho\, d\vartheta =$$

$$= \frac{4}{3}\sqrt{3}\left(\int_0^1 \rho^2\, d\rho\right)\left(\int_0^{\frac{\pi}{2}} \cos\vartheta\, d\vartheta\right) + \frac{4}{3}\sqrt{2}\left(\int_0^1 \rho^2\, d\rho\right)\left(\int_0^{\frac{\pi}{2}} \sin\vartheta\, d\vartheta\right) =$$

$$= \frac{4}{3}\sqrt{3}\left[\frac{1}{3}\rho^3\right]_0^1 \left[\sin\vartheta\right]_0^{\frac{\pi}{2}} + \frac{4}{3}\sqrt{2}\left[\frac{1}{3}\rho^3\right]_0^1 \left[-\cos\vartheta\right]_0^{\frac{\pi}{2}} = \frac{4}{9}\left(\sqrt{3}+\sqrt{2}\right).$$

# I – Double integrals

II

# TRIPLE INTEGRALS

## *Exercise 1*

Calculate the following triple integral over the specified set:

$$\int_{\Omega} xyz\,dx\,dy\,dz, \quad \Omega = \left\{(x,y,z) \in \mathbb{R}^3 : \; 0 \leq x \leq 1, \; 0 \leq y \leq 1, \; 0 \leq z \leq 1\right\}$$

Since the integrand function is the product of functions with separable variables, we simply have that:

$$\int_{\Omega} xyz\,dx\,dy\,dz = \left(\int_0^1 x\,dx\right)\left(\int_0^1 y\,dy\right)\left(\int_0^1 z\,dz\right) =$$

$$= \left[\frac{1}{2}x^2\right]_0^1 \left[\frac{1}{2}y^2\right]_0^1 \left[\frac{1}{2}z^2\right]_0^1 = \frac{1}{8}.$$

25

## II – Triple integrals

# *Exercise 2*

Calculate the following triple integral over the specified set:

$$\int_\Omega 2z\,dx\,dy\,dz, \quad \Omega = \left\{ (x,y,z) \in \mathbb{R}^3 : \ 2\sqrt{x^2+y^2} < z < x+2 \right\}$$

Integrating by wires parallel to the z axis, we have:

$$\int_\Omega 2z\,dx\,dy\,dz = 2\int_D \left[ \int_{2\sqrt{x^2+y^2}}^{x+2} z\,dz \right] dx\,dy =$$

$$= 2\int_D \left[ \frac{1}{2}z^2 \right]_{2\sqrt{x^2+y^2}}^{x+2} dx\,dy = \int_D \left[ (x+2)^2 - 4\left(x^2+y^2\right) \right] dx\,dy,$$

Having redefined the integration set as:

$$D = \left\{ (x,y) \in \mathbb{R}^2 : \ 2\sqrt{x^2+y^2} < x+2 \right\}.$$

It can be observed that:

$$2\sqrt{x^2+y^2} < x+2 \quad \Longleftrightarrow \quad \frac{\left(x-\frac{2}{3}\right)^2}{\frac{16}{9}} + \frac{y^2}{\frac{4}{3}} < 1.$$

So D is the set of points inside the ellipse of the above equation. Its graphical representation is given by:

## II – Triple integrals

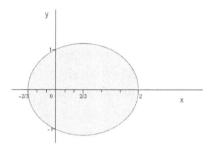

We pass into elliptical coordinates of the plane, i.e.:

$$\Phi : \begin{cases} x = \frac{2}{3} + \frac{4}{3}\rho\cos\vartheta \\ y = \frac{2}{3}\sqrt{3}\rho\sin\vartheta. \end{cases} \quad \rho \geq 0, \ 0 \leq \vartheta \leq 2\pi, \quad |\det J_\Phi(\rho,\vartheta)| = \frac{8}{9}\sqrt{3}\rho.$$

This results in:

$$(x,y) \in D \quad \Longleftrightarrow \quad \begin{cases} 0 \leq \rho < 1 \\ 0 \leq \vartheta < 2\pi. \end{cases}$$

$$D' = \left\{ (\rho,\vartheta) \in \mathbb{R}^2 : \ 0 \leq \rho < 1, \ 0 \leq \vartheta < 2\pi \right\}.$$

Substituting, we have:

$$\int_\Omega 2z\,dx\,dy\,dz = \int_D \left[ (x+2)^2 - 4\left(x^2+y^2\right) \right] dx\,dy =$$

$$= \int_D \left( 4 + 4x - 3x^2 - 4y^2 \right) dx\,dy = 3\int_D \left[ \frac{16}{9} - \left(x - \frac{2}{3}\right)^2 - \frac{4}{3}y^2 \right] dx\,dy =$$

$$= \frac{128}{27}\sqrt{3} \int_{D'} \left( \rho - \rho^3 \right) d\rho\,d\vartheta =$$

27

## II – Triple integrals

At this point we have separated the variables and therefore:

$$= \frac{128}{27} \sqrt{3} \left( \int_0^1 \left( \rho - \rho^3 \right) d\rho \right) \left( \int_0^{2\pi} d\vartheta \right) = \frac{256}{27} \sqrt{3} \pi \left[ \frac{1}{2}\rho^2 - \frac{1}{4}\rho^4 \right]_0^1 = \frac{64}{27} \sqrt{3} \pi.$$

## ***Exercise 3***

Calculate the following triple integral over the specified set:

$$\int_{\Omega} \frac{x^2}{x^2 + z^2} \, dx \, dy \, dz,$$

$$\Omega = \left\{ (x, y, z) \in \mathbb{R}^3 : 1 < x^2 + y^2 + z^2 < 2, \, x^2 - y^2 + z^2 < 0, \, y > 0 \right\}$$

To solve the integral, we pass in cylindrical coordinates:

$$\Phi : \begin{cases} x = \rho \cos \vartheta \\ y = y \\ z = \rho \sin \vartheta. \end{cases} \quad \rho \geq 0, \ 0 \leq \vartheta \leq 2\pi, \quad |\det J_{\Phi}(\rho, y, \vartheta)| = \rho.$$

Therefore:

## II – Triple integrals

$$(x, y, z) \in \Omega \quad \Longleftrightarrow \quad \begin{cases} 1 < \rho^2 + y^2 < 2 \\ \rho^2 - y^2 < 0 \\ y > 0. \end{cases}$$

$$\Omega' = \left\{ (\rho, y, \vartheta) \in \mathbb{R}^3 : \; 1 < \rho^2 + y^2 < 2, \; 0 \le \rho < y, \; 0 \le \vartheta < 2\pi \right\}.$$

Substituting:

$$\int_\Omega \frac{x^2}{x^2 + z^2} \, dx \, dy \, dz = \int_{\Omega'} \rho \cos^2 \vartheta \, d\rho \, dy \, d\vartheta =$$

Integrating for wires parallel to the theta axis:

$$= \int_D \left( \int_0^{2\pi} \rho \cos^2 \vartheta \, d\vartheta \right) d\rho \, dy = \left( \int_0^{2\pi} \cos^2 \vartheta \, d\vartheta \right) \left( \int_D \rho \, d\rho \, dy \right) =$$

$$= \left[ \frac{1}{2}(\vartheta + \sin \vartheta \cos \vartheta) \right]_0^{2\pi} \left( \int_D \rho \, d\rho \, dy \right) = \pi \left( \int_D \rho \, d\rho \, dy \right).$$

Where D is given by:

$$D = \left\{ (y, \rho) \in \mathbb{R}^2 : \; 1 < \rho^2 + y^2 < 2, \; 0 \le \rho < y \right\}.$$

To solve the integral it is better to make another passage of coordinates, passing to the polar ones:

$$\Psi : \begin{cases} y = r \cos \varphi \\ \rho = r \sin \varphi, \end{cases} \quad r \ge 0, \; 0 \le \varphi \le 2\pi, \quad |\det J_\Psi(r, \varphi)| = r.$$

## II – Triple integrals

Or:

$$(y, \rho) \in D \quad \Longleftrightarrow \quad \begin{cases} 1 < r < \sqrt{2} \\ 0 < \varphi < \frac{\pi}{4}. \end{cases}$$

$$D' = \left\{ (r, \varphi) \in \mathbb{R}^2 : \ 1 < r < \sqrt{2}, \ 0 < \varphi < \frac{\pi}{4} \right\}.$$

Substituting, we have:

$$\int_{\Omega} \frac{x^2}{x^2 + z^2} \, dx \, dy \, dz = \pi \left( \int_{D} \rho \, d\rho \, dy \right) = \pi \left( \int_{D'} r^2 \sin \varphi \, dr \, d\varphi \right) =$$

$$= \pi \left( \int_{0}^{\frac{\pi}{4}} \sin \varphi \, d\varphi \right) \left( \int_{1}^{\sqrt{2}} r^2 \, dr \right) = \pi \left[ -\cos \varphi \right]_{0}^{\frac{\pi}{4}} \left[ \frac{1}{3} r^3 \right]_{1}^{\sqrt{2}} = \frac{\pi}{6} \left( 5\sqrt{2} - 6 \right).$$

Having exploited the separation property of the variables in the third to last step.

## *Exercise 4*

Calculate the following triple integral over the specified set:

$$\int_{\Omega} \left( x^2 + y^2 + z^2 - 1 \right) \, dx \, dy \, dz,$$

$$\Omega = \left\{ (x, y, z) \in \mathbb{R}^3 : \ x^2 + y^2 + z^2 < 2, \ x^2 + y^2 < z \right\}$$

To solve the integral, we pass in cylindrical coordinates:

$$\Phi : \begin{cases} x = \rho \cos \vartheta \\ y = \rho \sin \vartheta \\ z = z, \end{cases} \quad \rho \geq 0, \ 0 \leq \vartheta \leq 2\pi, \quad |\det J_\Phi(\rho, y, \vartheta)| = \rho.$$

Therefore:

$$(x, y, z) \in \Omega \quad \Longleftrightarrow \quad \begin{cases} \rho^2 + z^2 < 2 \\ z > \rho^2 \\ 0 < \rho < 1. \end{cases}$$

$$\Omega' = \left\{ (\rho, \vartheta, z) \in \mathbb{R}^3 : \ 0 < \rho < 1, \ 0 \leq \vartheta < 2\pi, \ \rho^2 < z < \sqrt{2 - \rho^2} \right\}.$$

Substituting:

## II – Triple integrals

$$\int_{\Omega} \left( x^2 + y^2 + z^2 - 1 \right) dx\, dy\, dz = \int_{\Omega'} \left( \rho^2 + z^2 - 1 \right) \rho\, d\rho\, d\vartheta\, dz =$$

## Integrating for wires parallel to the zeta axis:

$$= \int_D \left[ \int_{\rho^2}^{\sqrt{2-\rho^2}} \left( \rho^2 + z^2 - 1 \right) \rho\, dz \right] d\rho\, d\vartheta =$$

$$= \int_D \rho \left[ \left( \rho^2 - 1 \right) z + \frac{1}{3} z^3 \right]_{\rho^2}^{\sqrt{2-\rho^2}} d\rho\, d\vartheta =$$

$$= \int_D \rho \left[ \left( \rho^2 - 1 \right) \sqrt{2-\rho^2} + \frac{1}{3} \left( 2-\rho^2 \right)^{\frac{3}{2}} - \left( \rho^2 - 1 \right) \rho^2 - \frac{1}{3}\rho^6 \right] d\rho\, d\vartheta =$$

$$= \int_D \left[ \rho^3 \sqrt{2-\rho^2} - \rho\sqrt{2-\rho^2} + \frac{1}{3}\rho \left( 2-\rho^2 \right)^{\frac{3}{2}} - \rho^5 + \rho^3 - \frac{1}{3}\rho^7 \right] d\rho\, d\vartheta,$$

## Where D is given by:

$$D = \left\{ (\rho, \vartheta) \in \mathbb{R}^2 : \ 0 < \rho < 1,\ 0 \le \vartheta < 2\pi \right\}.$$

## Proceeding with the accounts:

$$\int_{\Omega} \left( x^2 + y^2 + z^2 - 1 \right) dx\, dy\, dz =$$

$$= \left( \int_0^{2\pi} d\vartheta \right) \int_0^1 \left[ \rho^3\sqrt{2-\rho^2} - \rho\sqrt{2-\rho^2} + \frac{1}{3}\rho \left( 2-\rho^2 \right)^{\frac{3}{2}} - \rho^5 + \rho^3 - \frac{1}{3}\rho^7 \right] d\rho =$$

## Integrating the first addend by parts gives the result:

$$= 2\pi \left( -\frac{1}{3} + \frac{2}{3}\left[ -\frac{1}{5}\left( 2-\rho^2 \right)^{\frac{5}{2}} \right]_0^1 + \frac{1}{3} - \frac{2}{3}\sqrt{2} - \frac{1}{15} + \frac{4}{15}\sqrt{2} - \frac{1}{6} + \frac{1}{4} - \frac{1}{24} \right) =$$

$$= 2\pi \left( -\frac{2}{15} + \frac{8}{15}\sqrt{2} - \frac{2}{3}\sqrt{2} - \frac{1}{15} + \frac{4}{15}\sqrt{2} + \frac{1}{24} \right) = \pi \left( \frac{4}{15}\sqrt{2} - \frac{19}{60} \right).$$

# *Exercise 5*

Calculate the following triple integral over the specified set:

$$\int_{\Omega} (x+z)\,dx\,dy\,dz, \quad \Omega = \left\{(x,y,z) \in \mathbb{R}^3 : \ x > 0, \ y > 0, \ z > 0, \ x+y+z < 1\right\}$$

Integrating for wires parallel to the zeta axis:

$$\int_{\Omega}(x+z)\,dx\,dy\,dz = \int_D \left(\int_0^{1-x-y}(x+z)\,dz\right)dx\,dy = \int_D \left[xz + \frac{1}{2}z^2\right]_0^{1-x-y} dx\,dy =$$

$$= \int_D \left[x(1-x-y) + \frac{1}{2}(1-x-y)^2\right] dx\,dy.$$

Where D is given by:

$$D = \left\{(x,y) \in \mathbb{R}^2 : \ 0 < x < 1, \ 0 < y < 1-x\right\}$$

Proceeding with the accounts and noting that we can separate the variables:

$$\int_{\Omega}(x+z)\,dx\,dy\,dz = \int_0^1 \left(\int_0^{1-x}\left[x(1-x-y)+\frac{1}{2}(1-x-y)^2\right]dy\right)dx =$$

$$= \int_0^1 \left[x(1-x)y - \frac{1}{2}xy^2 - \frac{1}{6}(1-x-y)^3\right]_0^{1-x} dx =$$

$$= 2\pi\left(-\frac{1}{3}+\frac{2}{3}\left[-\frac{1}{5}\left(2-\rho^2\right)^{\frac{5}{2}}\right]_0^1 + \frac{1}{3} - \frac{2}{3}\sqrt{2} - \frac{1}{15} + \frac{4}{15}\sqrt{2} - \frac{1}{6} + \frac{1}{4} - \frac{1}{24}\right) =$$

$$= 2\pi\left(-\frac{2}{15}+\frac{8}{15}\sqrt{2}-\frac{2}{3}\sqrt{2}-\frac{1}{15}+\frac{4}{15}\sqrt{2}+\frac{1}{24}\right) = \pi\left(\frac{4}{15}\sqrt{2}-\frac{19}{60}\right).$$

II – Triple integrals

## *Exercise 6*

Calculate the following triple integral over the specified set:

$$\int_\Omega x|z|\,dx\,dy\,dz, \quad \Omega = \left\{ (x,y,z) \in \mathbb{R}^3 : \ \sqrt{x^2+z^2} < y < \frac{1}{2}x+3 \right\}$$

Both the function and the set have symmetry about the xy plane.
This results in:

$$\int_\Omega x|z|\,dx\,dy\,dz = 2\int_A xz\,dx\,dy\,dz,$$

Having defined A as:

$$A = \left\{ (x,y,z) \in \mathbb{R}^3 : \ \sqrt{x^2+z^2} < y < \frac{1}{2}x+3, \ z > 0 \right\}$$
$$= \left\{ (x,y,z) \in \mathbb{R}^3 : \ 0 < z < \sqrt{y^2-x^2}, \ |x| < y < \frac{1}{2}x+3 \right\}.$$

Integrating for wires parallel to the zeta axis:

34

## II – Triple integrals

$$\int_{\Omega} x|z|\,dx\,dy\,dz = 2\int_{A} xz\,dx\,dy\,dz = 2\int_{D}\left(\int_{0}^{\sqrt{y^2-x^2}} xz\,dz\right)dx\,dy =$$

$$= 2\int_{D}\left[\frac{1}{2}xz^2\right]_{0}^{\sqrt{y^2-x^2}}dx\,dy = \int_{D} x\left(y^2-x^2\right)dx\,dy,$$

Where D is given by:

$$D = \left\{(x,y)\in\mathbb{R}^2:\ |x| < y < \frac{1}{2}x+3\right\} = D_1 \cup D_2,$$

$$D_1 = \left\{(x,y)\in\mathbb{R}^2:\ -2 < x < 0,\ -x < y < \frac{1}{2}x+3\right\}$$

$$D_2 = \left\{(x,y)\in\mathbb{R}^2:\ 0 \le x < 6,\ x < y < \frac{1}{2}x+3\right\}.$$

We can separate the variables and get:

$$\int_{\Omega} x|z|\,dx\,dy\,dz = \int_{D} x\left(y^2-x^2\right)dx\,dy =$$

$$= \int_{D_1} x\left(y^2-x^2\right)dx\,dy + \int_{D_2} x\left(y^2-x^2\right)dx\,dy =$$

$$= \int_{-2}^{0}\left(\int_{-x}^{\frac{1}{2}x+3}\left(xy^2-x^3\right)dy\right)dx + \int_{0}^{6}\left(\int_{x}^{\frac{1}{2}x+3}\left(xy^2-x^3\right)dy\right)dx =$$

$$= \int_{-2}^{0}\left[\frac{1}{3}xy^3-x^3y\right]_{-x}^{\frac{1}{2}x+3}dx + \int_{0}^{6}\left[\frac{1}{3}xy^3-x^3y\right]_{x}^{\frac{1}{2}x+3}dx =$$

$$= \int_{-2}^{0}\left[\frac{1}{3}x\left(\frac{1}{2}x+3\right)^3-x^3\left(\frac{1}{2}x+3\right)+\frac{1}{3}x^4-x^4\right]dx +$$

$$+ \int_{0}^{6}\left[\frac{1}{3}x\left(\frac{1}{2}x+3\right)^3-x^3\left(\frac{1}{2}x+3\right)-\frac{1}{3}x^4+x^4\right]dx =$$

$$= \int_{-2}^{0}\left(-\frac{9}{8}x^4-\frac{9}{4}x^3+\frac{9}{2}x^2+9x\right)dx + \int_{0}^{6}\left(\frac{5}{24}x^4-\frac{9}{4}x^3+\frac{9}{2}x^2+9x\right)dx =$$

$$= \left[-\frac{9}{40}x^5-\frac{9}{16}x^4+\frac{3}{2}x^3+\frac{9}{2}x^2\right]_{-2}^{0} + \left[\frac{1}{24}x^5-\frac{9}{16}x^4+\frac{3}{2}x^3+\frac{9}{2}x^2\right]_{0}^{6} = \frac{384}{5}.$$

II – Triple integrals

## *Exercise 7*

Calculate the following triple integral over the specified set:

$$\int_\Omega 2x\,dx\,dy\,dz, \quad \Omega = \left\{(x,y,z) \in \mathbb{R}^3 : \; x > 0, \; 0 < y < 2z + 1, \; x^2 + y^2 + 4z^2 < 1\right\}$$

Integrating for wires parallel to the x axis:

$$\int_\Omega 2x\,dx\,dy\,dz = 2\int_D \left(\int_0^{\sqrt{1-y^2-4z^2}} x\,dx\right) dy\,dz =$$

$$= 2\int_D \left[\frac{1}{2}x^2\right]_0^{\sqrt{1-y^2-4z^2}} dy\,dz = \int_D \left(1 - y^2 - 4z^2\right) dy\,dz,$$

Where D is given by:

$$D = \left\{(z,y) \in \mathbb{R}^2 : \; 0 < y < 2z + 1, \; y^2 + 4z^2 < 1\right\} = D_1 \cup D_2,$$

$$D_1 = \left\{(z,y) \in \mathbb{R}^2 : \; -\frac{1}{2} < z \leq 0, \; 0 < y < 2z + 1\right\},$$

$$D_2 = \left\{(z,y) \in \mathbb{R}^2 : \; y^2 + 4z^2 < 1, \; y, z > 0\right\}.$$

Proceeding with the accounts:

36

## II – Triple integrals

$$\int_\Omega 2x \, dx \, dy \, dz = \int_D \left(1 - y^2 - 4z^2\right) dy \, dz =$$

$$= \int_{D_1} \left(1 - y^2 - 4z^2\right) dy \, dz + \int_{D_2} \left(1 - y^2 - 4z^2\right) dy \, dz.$$

The first integral is given by:

$$\int_{D_1} \left(1 - y^2 - 4z^2\right) dy \, dz = \int_{-\frac{1}{2}}^0 \left[\int_0^{2z+1} \left(1 - y^2 - 4z^2\right) dy\right] dz =$$

$$= \int_{-\frac{1}{2}}^0 \left[\left(1 - 4z^2\right)y - \frac{1}{3}y^3\right]_0^{2z+1} dz = \int_{-\frac{1}{2}}^0 \left(-\frac{32}{3}z^3 - 8z^2 + \frac{2}{3}\right) dz =$$

$$= \left[-\frac{8}{3}z^4 - \frac{8}{3}z^3 + \frac{2}{3}z\right]_{-\frac{1}{2}}^0 = \frac{1}{6}.$$

While for the second we pass in elliptical coordinates of the zy plane:

$$\Phi : \begin{cases} z = \frac{1}{2}\rho\cos\vartheta \\ y = \rho\sin\vartheta, \end{cases} \quad \rho \geq 0, \ 0 \leq \vartheta \leq 2\pi, \quad |\det J_\Phi(\rho, \vartheta)| = \frac{1}{2}\rho.$$

Or:

$$(z, y) \in D_2 \quad \Longleftrightarrow \quad \begin{cases} 0 < \rho < 1 \\ 0 < \vartheta < \frac{\pi}{2}. \end{cases}$$

$$D_2' = \left\{(\rho, \vartheta) \in \mathbb{R}^2 : \ 0 < \rho < 1, \ 0 < \vartheta < \frac{\pi}{2}\right\}.$$

## II – Triple integrals

Substituting, we have:

$$\int_{D_2} \left(1 - y^2 - 4z^2\right) dy\, dz = \int_{D_2'} \frac{1}{2}\rho \left(1 - \rho^2\right) d\rho\, d\vartheta =$$

$$= \left(\int_0^{\frac{\pi}{2}} d\vartheta\right) \left[\frac{1}{2}\int_0^1 \left(\rho - \rho^3\right) d\rho\right] = \frac{\pi}{4}\left[\frac{1}{2}\rho^2 - \frac{1}{4}\rho^4\right]_0^1 = \frac{\pi}{16}.$$

Definitely:

$$\int_\Omega 2x\, dx\, dy\, dz = \frac{\pi}{16} + \frac{1}{6}.$$

## *Exercise 8*

Calculate the following triple integral over the specified set:

$$\int_\Omega y\, dx\, dy\, dz,$$

$$\Omega = \left\{(x, y, z) \in \mathbb{R}^3 : \ x^2 + y^2 - 2x < 0, \ 0 < z < x, \ x^2 + y^2 < 1, \ y > 0\right\}$$

To solve the integral, we pass in cylindrical coordinates:

## II – Triple integrals

$$\Phi : \begin{cases} x = \rho \cos \vartheta \\ y = \rho \sin \vartheta \\ z = z, \end{cases} \quad \rho \geq 0, \ 0 \leq \vartheta \leq 2\pi, \quad |\det J_\Phi(\rho, y, \vartheta)| = \rho.$$

Therefore:

$$(x, y, z) \in \Omega \iff \begin{cases} 0 < \rho < 1 \\ \rho < 2\cos\vartheta \\ 0 < z < \rho\cos\vartheta \\ 0 < \vartheta < \frac{\pi}{2}. \end{cases}$$

$$\Omega' = \left\{ (\rho, \vartheta, z) \in \mathbb{R}^3 : \ 0 < \vartheta < \frac{\pi}{2}, \ 0 < \rho < 1, \ \rho < 2\cos\vartheta, \ 0 < z < \rho\cos\vartheta \right\}$$

Substituting:

$$\int_\Omega y \, dx \, dy \, dz = \int_{\Omega'} \rho \sin \vartheta \, d\rho \, dy \, d\vartheta =$$

Integrating for wires parallel to the zeta axis:

$$= \int_D \left( \int_0^{\rho\cos\vartheta} \rho^2 \sin\vartheta \, dz \right) d\rho \, d\vartheta = \int_D \rho^2 \sin\vartheta \Big[ z \Big]_0^{\rho\cos\vartheta} d\rho \, d\vartheta =$$

$$= \int_D \rho^3 \cos\vartheta \sin\vartheta \, d\rho \, d\vartheta,$$

Where D is given by the union of these two sets:

## II – Triple integrals

$$D_1 = \left\{ (\rho, \vartheta) \in \mathbb{R}^2 : \ 0 < \rho < 1, \ 0 < \vartheta < \frac{\pi}{3} \right\},$$

$$D_2 = \left\{ (\rho, \vartheta) \in \mathbb{R}^2 : \ \frac{\pi}{3} \leq \vartheta < \frac{\pi}{2}, \ 0 < \rho < 2\cos\vartheta \right\}$$

Proceeding with the accounts:

$$\int_\Omega y \, dx \, dy \, dz = \int_D \rho^3 \cos\vartheta \sin\vartheta \, d\rho \, d\vartheta =$$

$$= \int_{D_1} \rho^3 \cos\vartheta \sin\vartheta \, d\rho \, d\vartheta + \int_{D_2} \rho^3 \cos\vartheta \sin\vartheta \, d\rho \, d\vartheta =$$

By separating the variables, we get the result:

$$= \left( \int_0^1 \rho^3 \, d\rho \right) \left( \int_0^{\frac{\pi}{3}} \cos\vartheta \sin\vartheta \, d\vartheta \right) + \int_{\frac{\pi}{3}}^{\frac{\pi}{2}} \cos\vartheta \sin\vartheta \left[ \int_0^{2\cos\vartheta} \rho^3 \, d\rho \right] d\vartheta =$$

$$= \left[ \frac{1}{4}\rho^4 \right]_0^1 \left[ \frac{1}{2}\sin^2\vartheta \right]_0^{\frac{\pi}{3}} + \int_{\frac{\pi}{3}}^{\frac{\pi}{2}} \cos\vartheta \sin\vartheta \left[ \frac{1}{4}\rho^4 \right]_0^{2\cos\vartheta} d\vartheta =$$

$$= \frac{3}{32} + 4 \int_{\frac{\pi}{3}}^{\frac{\pi}{2}} \cos^5\vartheta \sin\vartheta \, d\vartheta = \frac{3}{32} + 4 \left[ -\frac{1}{6}\cos^6\vartheta \right]_{\frac{\pi}{3}}^{\frac{\pi}{2}} = \frac{5}{48}.$$

## II – Triple integrals

## *Exercise 9*

Calculate the following triple integral over the specified set:

$$\int_\Omega \frac{y^2}{x^2+y^2}\,dx\,dy\,dz, \quad \Omega = \left\{(x,y,z) \in \mathbb{R}^3: \ 1 < x^2+y^2 < 2x, \ 0 < z < \frac{x^2+y^2}{x^2}\right\}$$

Integrating for wires parallel to the zeta axis:

$$\int_\Omega \frac{y^2}{x^2+y^2}\,dx\,dy\,dz = \int_D \left(\int_0^{\frac{x^2+y^2}{x^2}} \frac{y^2}{x^2+y^2}\,dz\right) dx\,dy = \int_D \frac{y^2}{x^2}\,dx\,dy.$$

Where D is given by:

$$D = \left\{(x,y) \in \mathbb{R}^2: \ 1 < x^2 + y^2 < 2x\right\}.$$

We pass in polar coordinates of the xy plane:

$$\Phi: \begin{cases} x = \rho\cos\vartheta \\ y = \rho\sin\vartheta. \end{cases} \quad \rho \geq 0, \ -\pi \leq \vartheta \leq \pi, \quad |\det J_\Phi(\rho,\vartheta)| = \rho.$$

Therefore:

## II – Triple integrals

$$(x, y) \in D \iff \begin{cases} 1 < \rho < 2\cos\vartheta \\ -\dfrac{\pi}{3} < \vartheta < \dfrac{\pi}{3}. \end{cases}$$

$$D' = \left\{ (\rho, \vartheta) \in \mathbb{R}^2 : \quad -\frac{\pi}{3} < \vartheta < \frac{\pi}{3}, \ 1 < \rho < 2\cos\vartheta \right\}$$

Replacing and developing accounts:

$$\int_\Omega \frac{y^2}{x^2 + y^2}\, dx\, dy\, dz = \int_D \frac{y^2}{x^2}\, dx\, dy = \int_{D'} \frac{\sin^2\vartheta}{\cos^2\vartheta} \rho\, d\rho\, d\vartheta =$$

$$= \int_{-\frac{\pi}{3}}^{\frac{\pi}{3}} \left( \int_0^{2\cos\vartheta} \frac{\sin^2\vartheta}{\cos^2\vartheta} \rho\, d\rho \right) d\vartheta = \int_{-\frac{\pi}{3}}^{\frac{\pi}{3}} \frac{\sin^2\vartheta}{\cos^2\vartheta} \left[ \frac{1}{2}\rho^2 \right]_0^{2\cos\vartheta} d\vartheta =$$

$$= \frac{1}{2} \int_{-\frac{\pi}{3}}^{\frac{\pi}{3}} \frac{\sin^2\vartheta}{\cos^2\vartheta} \left( 4\cos^2\vartheta - 1 \right) d\vartheta = \frac{1}{2} \int_{-\frac{\pi}{3}}^{\frac{\pi}{3}} \left( 4\sin^2\vartheta - \tan^2\vartheta \right) d\vartheta =$$

We recall that:

$$\int \sin^2\vartheta\, d\vartheta = \frac{1}{2}(\vartheta - \cos\vartheta \sin\vartheta) + c$$

$$\int \tan^2\vartheta\, d\vartheta = \tan\vartheta - \vartheta + c.$$

We have the solution:

$$= \frac{1}{2} \left[ 2(\vartheta - \cos\vartheta \sin\vartheta) - \tan\vartheta + \vartheta \right]_{-\frac{\pi}{3}}^{\frac{\pi}{3}} = \pi - \frac{3}{2}\sqrt{3}.$$

## II – Triple integrals

# *Exercise 10*

Calculate the following triple integral over the specified set:

$$\int_{\Omega} 2z \, dx \, dy \, dz,$$

$$\Omega = \left\{ (x, y, z) \in \mathbb{R}^3 : \ 0 < y < x^2, \ x^2 - 2x + y^2 < 0, \ 0 < z < \sqrt{xy} \right\}$$

Integrating for wires parallel to the zeta axis:

$$\int_{\Omega} 2z \, dx \, dy \, dz = \int_D \left( \int_0^{\sqrt{xy}} 2z \, dz \right) dx \, dy = \int_D \left[ z^2 \right]_0^{\sqrt{xy}} dx \, dy = \int_D xy \, dx \, dy,$$

Where D is given by:

$$D = \left\{ (x, y) \in \mathbb{R}^2 : \ 0 < y < x^2, \ x^2 - 2x + y^2 < 0, \ x > 0 \right\}$$

And it is the union of these two sets:

$$D_1 = \left\{ (x, y) \in \mathbb{R}^2 : \ 0 < x \leq 1, \ 0 < y < x^2 \right\},$$

$$D_2 = \left\{ (x, y) \in \mathbb{R}^2 : \ 1 < x < 2, \ 0 < y < \sqrt{2x - x^2} \right\}.$$

Substituting, we have:

43

# II – Triple integrals

$$\int_\Omega 2z\,dx\,dy\,dz = \int_D xy\,dx\,dy = \int_{D_1} xy\,dx\,dy + \int_{D_2} xy\,dx\,dy =$$

$$= \int_0^1 \left( \int_0^{x^2} xy\,dy \right) dx + \int_1^2 \left( \int_0^{\sqrt{2x-x^2}} xy\,dy \right) dx =$$

$$= \int_0^1 \left[ \frac{1}{2}xy^2 \right]_0^{x^2} dx + \int_1^2 \left[ \frac{1}{2}xy^2 \right]_0^{\sqrt{2x-x^2}} dx = \frac{1}{2}\int_0^1 x^5\,dx + \frac{1}{2}\int_1^2 \left( 2x^2 - x^3 \right) dx =$$

$$= \frac{1}{2}\left[ \frac{1}{6}x^6 \right]_0^1 + \frac{1}{2}\left[ \frac{2}{3}x^3 - \frac{1}{4}x^4 \right]_1^2 = \frac{13}{24}.$$

## *Exercise 11*

Calculate the following triple integral over the specified set:

$$\int_\Omega \log \sqrt{x^2 + z^2}\,dx\,dy\,dz,$$

$$\Omega = \left\{ (x,y,z) \in \mathbb{R}^3 : \ 1 < x^2 + z^2 < e^2, \ z < x, \ 0 < y < \frac{1}{x^2 + z^2} \right\}$$

Integrating for wires parallel to the y axis:

$$\int_\Omega \log \sqrt{x^2 + z^2}\,dx\,dy\,dz = \int_D \left( \int_0^{\frac{1}{x^2+z^2}} \log \sqrt{x^2 + z^2}\,dy \right) dx\,dz =$$

$$= \int_D \frac{\log \sqrt{x^2 + z^2}}{x^2 + z^2}\,dx\,dz,$$

Where D is given by:

## II – Triple integrals

$$D = \left\{ (x, z) \in \mathbb{R}^2 : \ 1 < x^2 + z^2 < e^2, \ z < x \right\}$$

We pass in polar coordinates in the xz plane:

$$\Phi : \begin{cases} x = \rho \cos \vartheta \\ z = \rho \sin \vartheta, \end{cases} \quad \rho \geq 0, \ -\pi \leq \vartheta \leq \pi. \quad |\det J_\Phi(\rho, \vartheta)| = \rho.$$

Therefore:

$$(x, z) \in D \quad \Longleftrightarrow \quad \begin{cases} 1 < \rho < e \\ -\dfrac{3}{4}\pi < \vartheta < \dfrac{\pi}{4}. \end{cases}$$

$$D' = \left\{ (\rho, \vartheta) \in \mathbb{R}^2 : \ 1 < \rho < e, \ -\dfrac{3}{4}\pi < \vartheta < \dfrac{\pi}{4} \right\}$$

Substituting and separating the variables:

$$\int_\Omega \log \sqrt{x^2 + z^2} \, dx \, dy \, dz = \int_D \frac{\log \sqrt{x^2 + z^2}}{x^2 + z^2} \, dx \, dz = \int_{D'} \frac{\log \rho}{\rho} \, d\rho \, d\vartheta =$$

$$= \left( \int_{-\frac{3}{4}}^{\frac{\pi}{4}} d\vartheta \right) \left( \int_1^e \frac{\log \rho}{\rho} \, d\rho \right) = \pi \left[ \frac{1}{2} \log^2 \rho \right]_1^e = \frac{\pi}{2}.$$

## II – Triple integrals

## *Exercise 12*

Calculate the following triple integral over the specified set:

$$\int_{\Omega} y^2 |z| \, dx \, dy \, dz, \quad \Omega = \left\{ (x, y, z) \in \mathbb{R}^3 : \ 1 < x^2 + y^2 < 2x, \ 0 < z < \frac{2}{x} \right\}$$

Integrating for wires parallel to the z axis:

$$\int_{\Omega} y^2 |z| \, dx \, dy \, dz = \int_{\Omega} y^2 z \, dx \, dy \, dz = \int_D \left( \int_0^{\frac{2}{x}} y^2 z \, dz \right) dx \, dy =$$

$$= \int_D y^2 \left[ \frac{1}{2} z^2 \right]_0^{\frac{2}{x}} dx \, dy = 2 \int_D \frac{y^2}{x^2} \, dx \, dy.$$

Where D is given by:

$$D = \left\{ (x, z) \in \mathbb{R}^2 : \ 1 < x^2 + y^2 < 2x, \ x > 0 \right\}$$

Let's go to polar coordinates in the xy plane:

$$\Phi : \begin{cases} x = \rho \cos \vartheta \\ y = \rho \sin \vartheta, \end{cases} \quad \rho \geq 0, \ -\pi \leq \vartheta \leq \pi. \quad |\det J_\Phi(\rho, \vartheta)| = \rho.$$

Therefore:

## II – Triple integrals

$$(x, y) \in D \quad \Longleftrightarrow \quad \begin{cases} 1 < \rho < 2\cos\vartheta \\ -\frac{\pi}{3} < \vartheta < \frac{\pi}{3}. \end{cases}$$

$$D' = \left\{ (\rho, \vartheta) \in \mathbb{R}^2 : \quad -\frac{\pi}{3} < \vartheta < \frac{\pi}{3}, \ 1 < \rho < 2\cos\vartheta \right\}.$$

Substituting and separating the variables:

$$\int_\Omega y^2 |z| \, dx \, dy \, dz = 2 \int_D \frac{y^2}{x^2} \, dx \, dy = 2 \int_{D'} \rho \tan^2\vartheta \, d\rho \, d\vartheta =$$

$$= 2 \int_{-\frac{\pi}{3}}^{\frac{\pi}{3}} \left( \int_1^{2\cos\vartheta} \rho \tan^2\vartheta \, d\rho \right) d\vartheta = 2 \int_{-\frac{\pi}{3}}^{\frac{\pi}{3}} \tan^2\vartheta \left[ \frac{1}{2}\rho^2 \right]_1^{2\cos\vartheta} d\vartheta =$$

$$= 2 \int_{-\frac{\pi}{3}}^{\frac{\pi}{3}} \left( 4\sin^2\vartheta - \tan^2\vartheta \right) d\vartheta =$$

Recalling that:

$$\int \sin^2\vartheta \, d\vartheta = \frac{1}{2}(\vartheta - \cos\vartheta \sin\vartheta) + c$$

$$\int \tan^2\vartheta \, d\vartheta = \tan\vartheta - \vartheta + c$$

You get the result:

$$= 2 \left[ 2(\vartheta - \cos\vartheta \sin\vartheta) - \tan\vartheta + \vartheta \right]_{-\frac{\pi}{3}}^{\frac{\pi}{3}} = 2\pi - 3\sqrt{3}.$$

## *Exercise 13*

Calculate the following triple integral over the specified set:

$$\int_\Omega |z|\, dx\, dy\, dz, \quad \Omega = \left\{ (x,y,z) \in \mathbb{R}^3 : \ x^2 + y^2 < z^2 - 1, \ 2x^2 + y^2 + z^2 < 2 \right\}$$

We observe that the integrand function and the set are symmetric to the xy plane, therefore we have that:

$$\int_\Omega |z|\, dx\, dy\, dz = 2 \int_A z\, dx\, dy\, dz,$$

Where A is defined by:

$$\begin{aligned} A &= \left\{ (x,y,z) \in \mathbb{R}^3 : \ x^2 + y^2 < z^2 - 1, \ 2x^2 + y^2 + z^2 < 2, \ z > 0 \right\} \\ &= \left\{ (x,y,z) \in \mathbb{R}^3 : \ \sqrt{x^2 + y^2 + 1} < z < \sqrt{2 - 2x^2 - y^2} \right\}. \end{aligned}$$

Integrating for wires parallel to the z axis:

## II – Triple integrals

$$\int_\Omega |z|\, dx\, dy\, dz = 2 \int_A z\, dx\, dy\, dz = 2 \int_D \left( \int_{\sqrt{x^2+y^2+1}}^{\sqrt{2-2x^2-y^2}} z\, dz \right) dx\, dy =$$

$$= 2 \int_D \left[ \frac{1}{2} z^2 \right]_{\sqrt{x^2+y^2+1}}^{\sqrt{2-2x^2-y^2}} dx\, dy = \int_D \left( 1 - 3x^2 - 2y^2 \right) dx\, dy.$$

## Where D is given by:

$$D = \left\{ (x,y) \in \mathbb{R}^2 : \ \sqrt{x^2+y^2+1} < \sqrt{2-2x^2-y^2} \right\}$$

$$= \left\{ (x,y) \in \mathbb{R}^2 : \ 3x^2 + 2y^2 < 1 \right\}.$$

## We pass to elliptical coordinates in the xy plane:

$$\Phi : \begin{cases} x = \frac{\sqrt{3}}{3} \rho \cos \vartheta \\ y = \frac{\sqrt{2}}{2} \rho \sin \vartheta, \end{cases} \quad \rho \geq 0,\ 0 \leq \vartheta \leq 2\pi. \quad |\det J_\Phi(\rho, \vartheta)| = \frac{\sqrt{6}}{6}\rho.$$

## Therefore:

$$(x,y) \in D \iff \begin{cases} 0 \leq \rho < 1 \\ 0 \leq \vartheta < 2\pi. \end{cases}$$

$$D' = \left\{ (\rho, \vartheta) \in \mathbb{R}^2 : \ 0 \leq \rho < 1,\ 0 \leq \vartheta < 2\pi \right\}$$

## Substituting and separating the variables:

$$\int_\Omega |z|\, dx\, dy\, dz = \int_D \left( 1 - 3x^2 - 2y^2 \right) dx\, dy = \frac{\sqrt{6}}{6} \int_{D'} \rho \left( 1 - \rho^2 \right) d\rho\, d\vartheta =$$

49

## II – Triple integrals

$$= \frac{\sqrt{6}}{3}\pi \int_0^1 \left(\rho - \rho^3\right) d\rho = \frac{\sqrt{6}}{3}\pi \left[\frac{1}{2}\rho^2 - \frac{1}{4}\rho^4\right]_0^1 = \frac{\sqrt{6}}{12}\pi.$$

# *Exercise 14*

Calculate the following triple integral over the specified set:

$$\int_\Omega x^2|y|\, dx\, dy\, dz, \quad \Omega = \left\{(x,y,z) \in \mathbb{R}^3 : \; x^2 + y^2 + z^2 < 1, \; 0 < z < x + 1\right\}$$

We observe that the integrand function and the set are symmetric on the xz plane, therefore we have that:

$$\int_\Omega x^2|y|\, dx\, dy\, dz = 2 \int_A x^2 y\, dx\, dy\, dz,$$

Where A is defined by:

$$
\begin{aligned}
A &= \left\{(x,y,z) \in \mathbb{R}^3 : \; x^2 + y^2 + z^2 < 1, \; 0 < z < x + 1, \; y > 0\right\} \\
&= \left\{(x,y,z) \in \mathbb{R}^3 : \; 0 < y < \sqrt{1 - x^2 - z^2}, \; 0 < z < x + 1\right\}.
\end{aligned}
$$

Integrating for wires parallel to the y axis:

## II – Triple integrals

$$\int_\Omega x^2 |y| \, dx \, dy \, dz = 2 \int_A x^2 y \, dx \, dy \, dz = 2 \int_D \left( \int_0^{\sqrt{1-x^2-z^2}} x^2 y \, dy \right) dx \, dz =$$

$$= 2 \int_D x^2 \left[ \frac{1}{2} y^2 \right]_0^{\sqrt{1-x^2-z^2}} dx \, dz = \int_D x^2 \left( 1 - x^2 - z^2 \right) dx \, dz,$$

$$D = \left\{ (x,z) \in \mathbb{R}^2 : \ x^2 + z^2 < 1, \ 0 < z < x + 1 \right\} = D_1 \cup D_2,$$

$$D_1 = \left\{ (x,z) \in \mathbb{R}^2 : \ -1 < x \le 0, \ 0 < z < x + 1 \right\},$$

$$D_2 = \left\{ (x,z) \in \mathbb{R}^2 : \ x^2 + z^2 < 1, \ x,z > 0 \right\}.$$

## Substituting, we have:

$$\int_\Omega x^2 |y| \, dx \, dy \, dz = \int_D x^2 \left( 1 - x^2 - z^2 \right) dx \, dz =$$

$$= \int_{D_1} x^2 \left( 1 - x^2 - z^2 \right) dx \, dz + \int_{D_2} x^2 \left( 1 - x^2 - z^2 \right) dx \, dz.$$

## The first and second integrals develop as follows:

$$\int_{D_1} x^2 \left( 1 - x^2 - z^2 \right) dx \, dz = \int_{-1}^0 \left[ \int_0^{x+1} x^2 \left( 1 - x^2 - z^2 \right) dz \right] dx =$$

$$= \int_{-1}^0 x^2 \left[ \left( 1 - x^2 \right) z - \frac{1}{3} z^3 \right]_0^{x+1} dx = \int_{-1}^0 x^2 \left[ \left( 1 - x^2 \right) (x+1) - \frac{1}{3} (x+1)^3 \right] dx =$$

$$= \int_{-1}^0 \left( -\frac{4}{3} x^5 - 2x^4 + \frac{2}{3} x^2 \right) dx = \left[ -\frac{2}{9} x^6 - \frac{2}{5} x^5 + \frac{2}{9} x^3 \right]_{-1}^0 = \frac{2}{45}.$$

$$\Phi : \begin{cases} x = \rho \cos \vartheta \\ z = \rho \sin \vartheta, \end{cases} \quad \rho \ge 0, \ 0 \le \vartheta \le 2\pi, \quad |\det J_\Phi(\rho,\vartheta)| = \rho.$$

$$(x,z) \in D_2 \quad \Longleftrightarrow \quad \begin{cases} 0 < \rho < 1 \\ 0 < \vartheta < \frac{\pi}{2}. \end{cases}$$

## II – Triple integrals

$$D'_2 = \left\{ (\rho, \vartheta) \in \mathbb{R}^2 : \ 0 < \rho < 1, \ 0 < \vartheta < \frac{\pi}{2} \right\}.$$

Substituting and separating the variables:

$$\int_{D_2} x^2 \left( 1 - x^2 - z^2 \right) dx \, dz = \int_{D'_2} \rho^3 \cos^2 \vartheta \left( 1 - \rho^2 \right) d\rho \, d\vartheta =$$

$$= \left( \int_0^{\frac{\pi}{2}} \cos^2 \vartheta \, d\vartheta \right) \left[ \int_0^1 \left( \rho^3 - \rho^5 \right) d\rho \right] = \left[ \frac{1}{2}(\vartheta + \sin \vartheta \cos \vartheta) \right]_0^{\frac{\pi}{2}} \left[ \frac{1}{4}\rho^4 - \frac{1}{6}\rho^6 \right]_0^1 = \frac{\pi}{48}.$$

The final result is given by:

$$\int_\Omega x^2 |y| \, dx \, dy \, dz = \frac{\pi}{48} + \frac{2}{45}.$$

# II – Triple integrals